# Preface

Welcome to Science Olympiad All Chapter Wise Story Book Class 1. This book has been created to make science an enjoyable and meaningful experience for young learners. We believe that children in their formative years learn best through stories, hands-on activities, and relatable examples. Keeping this in mind, we have designed this book to introduce scientific concepts in a way that captures their imagination and encourages curiosity.

Each chapter covers a key topic from the Science Olympiad syllabus, ensuring a focused and thorough preparation for the competition. However, this book goes beyond exam preparation. By integrating science into everyday scenarios and weaving it into engaging narratives, we aim to foster a love for exploration and discovery that lasts a lifetime.

To enhance learning, each chapter includes thoughtfully designed questions, illustrations, and activities that encourage students to think critically and creatively. These features are not only intended to reinforce understanding but also to inspire students to ask questions, seek answers, and explore the wonders of the natural world.

To parents and teachers, we invite you to join your children and students on this learning journey. Read along with them, discuss the stories, and explore the activities together. Your guidance and encouragement will make this experience even more impactful and enjoyable.

We hope this book will serve as a companion in your child's educational journey, helping them excel academically while instilling a passion for science. Thank you for choosing this book, and we wish you and your young learners success and joy as you embark on this adventure into the fascinating world of science.

# Table of Contents

## Chapter 1 - Weather and Seasons

Once upon a time, in a little town called Weatherville, there lived three best friends named Sunny, Rainy, and Cloudy. These friends were very special because they controlled the weather in Weatherville.

One bright morning, Sunny woke up early and decided to visit his friend, Rainy. "Good morning, Rainy!" said Sunny with a big smile. "Let's play today!" Rainy smiled back and said, "That sounds like fun, but I have a surprise for everyone first."

Sunny watched as Rainy spread his magic, and suddenly, tiny raindrops began to fall from the sky. The flowers and trees in Weatherville were so happy to drink the fresh rainwater. The kids grabbed their colorful umbrellas and rain boots to splash in the puddles.

After a while, Cloudy floated over. "Hi, Sunny! Hi, Rainy!" Cloudy said. "Why don’t we take a break now?" Cloudy waved his fluffy hands, and the sky filled with soft, white clouds. The rain stopped, and the air felt cool and fresh. The children put away their umbrellas and decided to play in the cool shade.

As the day went on, the weather changed again. The clouds drifted away, and the sun shone brightly. Sunny was back, and it was a warm afternoon. The children took off their sweaters and played in the sunshine, enjoying the warm weather.

Later that evening, as the sun began to set, the friends gathered to talk about the day. "Today was perfect," said Sunny. "Everyone got to enjoy a little bit of everything!" Rainy nodded and added, "Yes, the flowers are happy, and so are the children!" Cloudy agreed, "It’s fun to see how different weather can make everyone feel."

The three friends knew that each of them was important, and together, they made Weatherville a wonderful place to live. And so, Sunny, Rainy, and Cloudy continued to bring their special weather to the town, helping everyone enjoy each new day.

## Activities

1. **Weather Dance:** Ask the children to act out different types of weather while listening to music. For example, they can jump like raindrops during a rainy song, spin like the wind during a windy song, or stretch out like sun rays during a sunny song.

2. **Weather Chart:** Create a weather chart in the classroom. Every morning, ask the children to look outside and decide what the weather is like. They can place a picture of the sun, rain, wind, or clouds on the chart.

3. **Seasonal Dressing:** Bring different types of clothing (like a sweater, raincoat, sunglasses, etc.) to the class. Ask the children to choose which clothes they would wear for each type of weather.

4. **Storytime Discussion:** After reading the story, ask the children which type of weather they like the most and why. Let them share their experiences of playing in the rain, snow, or sunshine.

5. **Draw Your Favorite Weather:** Provide drawing materials and ask the children to draw a picture of their favorite type of weather. Encourage them to include details like what they wear, what they do, and how the sky looks.

## Questions

1. What are the names of the three friends in Weatherville?

   a) Sun, Moon, and Stars

   b) Sunny, Rainy, and Cloudy

   c) Day, Night, and Noon

   d) Windy, Stormy, and Snowy

2. What did Rainy do when Sunny wanted to play?

   a) Made it sunny

   b) Made it rain

   c) Made it windy

   d) Made it snow

3. How did the flowers and trees feel when it rained?

   a) Happy

   b) Sad

   c) Angry

   d) Sleepy

4. What did the children use when it started raining?

a) Sunglasses

b) Rain boots and umbrellas

c) Hats and gloves

d) Sandals and sunglasses

5. Who brought the cool shade in the story?

a) Sunny

b) Rainy

c) Cloudy

d) Windy

6. What happened after Cloudy's shade?

a) It rained again

b) The sun shone brightly

c) It became nighttime

d) It snowed

7. What do the children do when the sun shines brightly?

a) Put on sweaters

b) Play in the sunshine

c) Go inside

d) Look for clouds

8. How did Sunny, Rainy, and Cloudy feel about their day?

a) Sad

b) Angry

c) Happy

d) Sleepy

9. What does Cloudy do to the sky?

a) Fills it with clouds

b) Clears the clouds away

c) Makes it rain

d) Makes it snow

10. Why is each type of weather important in Weatherville?

a) Because it makes the town a fun place to live

b) Because it makes everyone stay inside

c) Because it always snows

d) Because it's always sunny

## Chapter 2 - Food and Water

Benny was a little boy who loved to play and explore. One sunny day, his mother said, "Benny, let’s go on a picnic! We’ll pack some delicious food and plenty of water." Benny was excited and quickly helped his mother pack a basket with fruits, vegetables, sandwiches, and a big bottle of water.

They went to a beautiful park near a sparkling lake. Benny’s mother spread out a blanket, and they sat down to enjoy their picnic. As they began to eat, a friendly bird flew over and said, "Hello, Benny! Did you know that food and water are very important for your body?"

Benny looked at the bird in surprise. "Really? Why?" he asked.

The bird chirped happily, "Food gives you energy to play and grow strong. Different foods help different parts of your body. Fruits and vegetables keep you healthy, grains give you energy, and proteins like eggs and cheese help your muscles grow!"

Benny looked at his plate. He had apples, carrots, bread, and a piece of cheese. "So, all these foods are helping me in different ways?" Benny asked.

"Yes, exactly!" said the bird. "And don’t forget water. Water is just as important as food. It keeps your body cool, helps you digest your food, and keeps your skin and hair healthy."

Benny took a big gulp of water and smiled. "I feel so refreshed!"

The bird continued, "There are many sources of water, like rivers, lakes, and rain. It’s important to drink clean water every day. And remember, always eat a variety of foods to stay healthy."

Benny's mother nodded and said, "That's why we brought different kinds of food today, Benny. Eating a little bit of everything helps your body stay strong and healthy."

Benny felt proud of his healthy picnic. After they finished eating, they played games and enjoyed the sunny day. Benny knew that the food and water he had enjoyed would give him the energy to play all day long.

Before they left, the bird gave Benny one last piece of advice. "Remember, Benny, eat slowly, chew your food well, and drink plenty of water throughout the day. Your body will thank you!"

Benny promised to remember the bird's advice. As they packed up their picnic and headed home, Benny felt happy and healthy, knowing that he was taking good care of his body.

## Activities

1. **Healthy Plate Activity:** Provide children with pictures of different foods (fruits, vegetables, grains, proteins) and ask them to create a healthy plate by gluing the pictures onto a paper plate.

2. **Water Cycle Experiment:** Demonstrate the water cycle with a simple experiment. Use a bowl of water to represent a lake, cover it with plastic wrap, and place it in the sun to show how water evaporates and then condenses, just like rain.

3. **Food Sorting Game:** Provide a mix of food items (real or pictures) and ask the children to sort them into categories: fruits, vegetables, grains, and proteins.

4. **Hydration Station:** Set up a "Hydration Station" where children can fill up their water bottles and learn about the importance of drinking water throughout the day.

5. **Healthy Eating Habits Chart:** Create a chart where children can track their daily intake of different food groups and water. Encourage them to eat a variety of foods and drink plenty of water.

## Questions

1. Why are food and water important for your body?

   a) To help you sleep

   b) To give you energy and keep you healthy

   c) To make you laugh

   d) To help you sing

2. What types of food did Benny eat at the picnic?

   a) Fruits, Vegetables, Grains, Proteins

   b) Candy, Chips, Soda

   c) Ice Cream, Cookies, Cake

   d) Pizza, Burgers, Fries

3. What does water help your body do?

   a) Keep cool and digest food

   b) Make you run faster

   c) Help you jump higher

   d) Help you sleep longer

4. Where can we get water from?

a) Rivers, Lakes, Rain

b) Clouds, Stars, Moon

c) Trees, Flowers, Grass

d) Shoes, Hats, Coats

5. What should you do when you eat food?

a) Eat slowly and chew well

b) Eat quickly without chewing

c) Talk loudly and eat

d) Play while eating

6. Why should you drink clean water every day?

a) To stay healthy

b) To get dirty

c) To play games

d) To jump higher

7. What helps your muscles grow?

a) Proteins like eggs and cheese

b) Ice cream and cookies

c) Soda and chips

d) Candy and chocolate

8. Why is it important to eat a variety of foods?

a) To stay strong and healthy

b) To eat more candy

c) To stay awake all night

d) To play more games

9. What should you do if you're feeling thirsty?

a) Drink water

b) Eat candy

c) Go to sleep

d) Run fast

10. What did Benny promise to do after his picnic?

a) Eat slowly and drink plenty of water

b) Eat lots of candy

c) Stop eating vegetables

d) Drink soda instead of water

## Chapter 3 - Travel and Safety

One sunny morning, Tia woke up excited because she was going to visit Safety Town with her friends. Safety Town was a special place where children could learn all about traveling and staying safe. Tia quickly got dressed and met her friends at the bus stop.

As they boarded the bus, their teacher, Ms. Green, reminded them to hold onto the railings and sit quietly in their seats. "Remember, we must always be careful when traveling," she said with a smile. Tia nodded and looked out the window as the bus took them to Safety Town.

When they arrived, the children were greeted by Officer Sam, a friendly police officer. "Welcome to Safety Town!" Officer Sam said. "Today, we're going to learn all about different ways to travel and how to stay safe."

First, Officer Sam showed them the different ways people travel. "Some people walk, others ride bicycles, and some take cars, buses, trains, or even airplanes!" he explained. Tia imagined herself flying high in the sky in an airplane, seeing the world from above.

Next, Officer Sam took them to the Road Safety area. There was a pretend road with traffic signals and zebra crossings. "It's very important to follow road safety rules," said Officer Sam. "Always stop, look, and listen before crossing the road. And remember the colors of the traffic lights: Red means stop, yellow means slow down, and green means go!"

Tia and her friends practiced crossing the pretend road. They waited for the green light, looked both ways, and walked carefully across the zebra crossing. "Good job!" said Officer Sam. "You're all being very safe!"

Then, they moved to the Safe Travel Tips section. Here, they learned about wearing seat belts in cars, helmets when cycling, and always staying seated on the bus. "These are important rules to keep you safe while traveling," Officer Sam explained.

Finally, Officer Sam talked about safety at home and school. "At home, never touch hot stoves or sharp objects, and always ask an adult for help if you're unsure. At school, walk, don't run in the hallways, and be careful on the playground."

After a fun and informative day at Safety Town, Tia felt confident about staying safe wherever she went. "Thank you, Officer Sam!" Tia said as they got back on the bus. "I've learned so much about travel and safety!"

On the way home, Tia thought about all the things she had learned. She knew that by following the safety rules, she could always have fun adventures and stay safe at the same time.

## Activities

1. **Traffic Signal Game:** Use red, yellow, and green cards. The teacher shows a card, and the children must act according to the signal: Stop (red), Slow down (yellow), and Go (green).

2. **Cross the Road Practice:** Set up a pretend road in the classroom with a zebra crossing. Have the children practice looking both ways and crossing the road safely when the "traffic light" (teacher) turns green.

3. **Safety Equipment Show and Tell:** Bring items like helmets, seat belts, and reflectors. Let the children touch and wear them while explaining why each item is important for safety.

4. **Roleplaying:** Set up different scenarios where children practice safety rules, like crossing the road, sitting on a bus, or playing safely at home.

5. **Safety Posters:** Provide drawing materials for children to create posters about road safety, safe travel tips, or safety at home and school. Display the posters in the classroom.

## Questions

1. Where did Tia and her friends go on their adventure?

   a) Fun Town

   b) Safety Town

   c) Adventure Land

   d) Play Park

2. Who greeted the children at Safety Town?

   a) Firefighter Tom

   b) Officer Sam

   c) Doctor Lily

   d) Teacher Green

3. What are some ways people travel?

   a) Walking, Cycling, Car, Bus, Train, Airplane

   b) Skipping, Dancing, Flying like a bird

   c) Running, Hopping, Jumping

d) Sleeping, Sitting, Standing

4. What does a red traffic light mean?

a) Go

b) Stop

c) Run

d) Dance

5. What should you do before crossing the road?

a) Run fast

b) Stop, look, and listen

c) Jump up and down

d) Close your eyes

6. What should you wear when cycling to stay safe?

a) Sunglasses

b) Helmet

c) Scarf

d) Boots

7. Why is it important to wear a seat belt in a car?

a) To look cool

b) To stay safe in case of sudden stops

c) To be taller

d) To sleep better

8. What should you never touch at home without an adult's help?

a) Toys

b) Hot stoves or sharp objects

c) Pillows

d) Books

9. What should you do in the school hallways?

a) Run as fast as you can

b) Walk carefully

c) Jump up and down

d) Sit on the floor

10. What did Tia learn about by visiting Safety Town?

a) How to play games

b) How to stay safe while traveling and at home or school

c) How to build a house

d) How to cook food

## Chapter 4 - Human Body

Once upon a time, in a little town, there lived a curious boy named Sammy. Sammy loved to explore and learn new things. One day, while playing in his room, he found a beautiful, old mirror. As he looked into the mirror, something magical happened. The mirror spoke!

"Hello, Sammy! I'm the Talking Mirror, and I can teach you all about your body!" Sammy was amazed and excited. "Wow! Can you really?" he asked.

"Yes, I can!" replied the mirror. "Let's start with the parts of your body. Look at your head it's where your brain is, and it helps you think and learn. Your eyes, ears, nose, and mouth are all on your head, and they help you see, hear, smell, and taste."

Sammy touched his head and smiled. "That's right! My head is very important!"

The mirror continued, "Next, look at your arms. Your arms help you pick things up, hug your friends, and wave hello! And don't forget your hands and fingers they help you hold things and write."

Sammy wiggled his fingers and laughed. "My hands are so useful!"

"Now, let's move to your legs," the mirror said. "Your legs help you walk, run, jump, and play. Your feet and toes help you balance and move around."

Sammy jumped up and down. "I love running and playing with my legs!"

The mirror then explained, "But it's not just about having body parts—we need to take care of them too! You should brush your teeth every day, take baths to keep your skin clean, and do exercises to keep your body strong and healthy."

Sammy nodded, "I brush my teeth every morning and night. And I love playing outside it's like exercise!"

"That's wonderful, Sammy!" said the mirror. "But did you know your body also has special helpers called senses? Your eyes help you see beautiful things, your ears help you hear music and voices, your nose helps you smell flowers and food, your tongue helps you taste yummy treats, and your skin helps you feel warm and soft things."

Sammy closed his eyes and imagined all the wonderful things his senses could do. "My senses are like little superpowers!"

"Exactly, Sammy!" the mirror said. "Your body is amazing, and by taking care of it and using your senses, you can enjoy everything around you."

From that day on, Sammy remembered everything the Talking Mirror had taught him. He took good care of his body, used his senses wisely, and always had fun exploring the world around him.

## Activities

1. **Body Parts Song:** Sing a song that names different parts of the body (like "Head, Shoulders, Knees, and Toes") and ask the children to point to each part as they sing.

2. **Senses Exploration:** Set up five stations, each focusing on one of the senses. For example, smelling flowers, tasting fruit, touching soft toys, looking at colorful pictures, and listening to different sounds.

3. **Body Parts Puzzle:** Create a simple puzzle of the human body, where children can place different body parts (head, arms, legs, etc.) in the correct spots.

4. **Healthy Habits Chart:** Help children create a daily chart where they can check off healthy habits like brushing teeth, bathing, and exercising.

5. **Sense Detective:** Play a game where children guess what they are smelling, hearing, touching, tasting, or seeing with their eyes closed or using a blindfold.

## Questions

1. What is on top of your body and helps you think?

   a) Arm

   b) Leg

   c) Head

   d) Foot

2. What body part helps you pick things up and hug your friends?

   a) Eyes

   b) Ears

   c) Arms

   d) Nose

3. Which body parts help you walk, run, and jump?

   a) Fingers

   b) Legs

   c) Ears

   d) Mouth

4. What should you do every day to keep your teeth clean?

   a) Jump

   b) Brush them

c) Clap your hands

d) Run fast

5. What does your tongue help you do?

a) See things

b) Hear sounds

c) Taste food

d) Smell flowers

6. What should you do to keep your skin clean?

a) Sing a song

b) Take a bath

c) Jump up and down

d) Sleep all day

7. Which sense helps you hear music and voices?

a) Seeing

b) Hearing

c) Smelling

d) Touching

8. What do your eyes help you do?

a) Taste food

b) Hear sounds

c) See things

d) Smell flowers

9. Why is it important to do exercises?

a) To keep your body strong and healthy

b) To make your clothes dirty

c) To stay awake all night

d) To eat more candy

10. What are the five senses that help you enjoy the world around you?

a) Jumping, running, sleeping, eating, and talking

b) Seeing, hearing, tasting, smelling, and touching

c) Laughing, crying, sleeping, dancing, and singing

d) Sitting, standing, swimming, biking, and flying

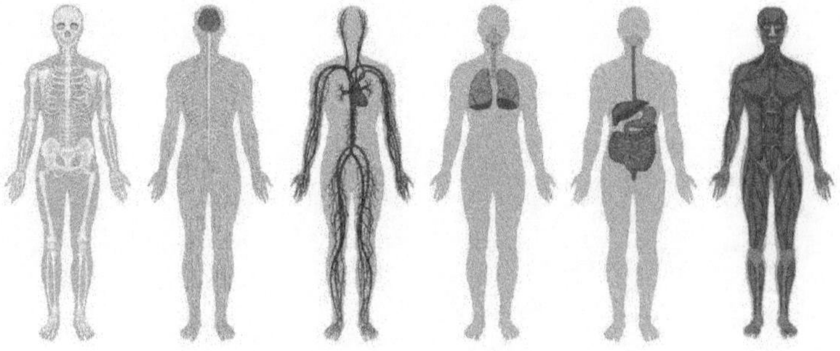

## Chapter 5 - Plants

In a small village surrounded by hills, there was a little girl named Lily. She loved spending time outdoors, especially in her grandmother's garden. The garden was full of colorful flowers, tall trees, and small, green plants. One day, as Lily was watering the plants, she noticed something sparkling in the soil. Curious, she dug a little and found a tiny, glowing seed.

Excited, Lily ran to her grandmother and asked, "Grandma, what is this glowing seed?" Her grandmother smiled and said, "That is a Magic Seed, Lily. If you plant it in the right spot and take good care of it, something wonderful will grow."

Lily couldn't wait to plant the Magic Seed. She carefully placed it in the soil, covered it up, and watered it gently. Every day, she visited the spot, watering the seed and making sure it got plenty of sunlight.

A few days later, Lily saw a little green shoot coming out of the soil. "Grandma, it's growing!" she shouted with joy. Her grandmother came over and said, "Yes, Lily, that's the start of a plant. Every plant has different parts that help it grow."

Lily listened carefully as her grandmother explained. "The roots grow underground. They drink water and help the plant stay strong. The stem holds the plant up and carries water from the roots to the leaves. The leaves catch sunlight and help make food for the plant. And some plants grow beautiful flowers, which can turn into fruits."

As the days passed, the little plant grew taller. Lily noticed that it had small, delicate leaves and a strong stem. One morning, she saw something new—a tiny flower bud at the top of the stem. "Grandma, look! A flower is going to bloom!" Lily exclaimed.

Her grandmother smiled and said, "That's right, Lily. Plants are very important. They give us food like fruits and vegetables, they make the air fresh and clean, and they make the world more beautiful with their flowers."

Lily nodded, feeling proud of her little plant. "I'll take good care of you, little plant," she said softly. She watered it every day, made sure it got sunlight, and even talked to it.

The plant continued to grow, and soon the garden was filled with its beautiful flowers. Lily's garden became the most magical place in the village, full of life and color.

## Activities

1. **Plant a Seed:** Give each child a small pot, some soil, and a seed. Let them plant their own seed, water it, and watch it grow over time. Encourage them to observe the roots, stem, leaves, and flowers as the plant develops.

2. **Parts of a Plant Puzzle:** Create a puzzle with the different parts of a plant (roots, stem, leaves, flowers). Have the children assemble the puzzle and label each part.

3. **Types of Plants Scavenger Hunt:** Take the children on a nature walk or show pictures of different plants. Ask them to find examples of trees, shrubs, and herbs.

4. **Caring for Plants Chart:** Create a chart where children can check off tasks like watering, giving sunlight, and keeping the plants safe from harm.

5. **Coloring Flowers:** Provide coloring sheets with different types of plants and flowers. Let the children color them and learn the names of each plant.

## Questions:

1. What did Lily find in her grandmother's garden?

   a) A shiny rock

   b) A Magic Seed

   c) A toy

   d) A coin

2. What does a plant's stem do?

   a) Holds the plant up and carries water

   b) Grows flowers

   c) Makes seeds

   d) Makes the plant fly

3. What part of the plant grows underground?

   a) Leaves

   b) Flowers

   c) Roots

   d) Stem

4. What do leaves help the plant do?

   a) Catch sunlight and make food

   b) Hold the plant up

c) Grow flowers

d) Make the plant smell good

5. Which of these is NOT a part of a plant?

a) Roots

b) Stem

c) Leaves

d) House

6. What types of plants did Lily's grandmother talk about?

a) Trees, Shrubs, and Herbs

b) Cars, Bikes, and Trucks

c) Rocks, Sand, and Water

d) Toys, Books, and Games

7. Why are plants important?

a) They make toys

b) They give us food, air, and beauty

c) They make noise

d) They run fast

8. How did Lily take care of her plant?

a) By ignoring it

b) By giving it water, sunlight, and love

c) By pulling it out of the soil

d) By covering it with rocks

9. What did the plant in Lily's garden grow?

a) A treehouse

b) Flowers

c) A car

d) A bird

10. What should you do to take care of plants?

a) Water them and give them sunlight

b) Never look at them

c) Yell at them

d) Forget about them

## Chapter 6 - Clothes

In a small town called Dressville, there lived a little boy named Timmy. Timmy loved playing outside, but he always had trouble deciding what to wear. One day, while playing in his room, he found an old wardrobe that he had never seen before. It was big, made of dark wood, and had shiny golden handles.

Curious, Timmy opened the wardrobe, and to his surprise, it started to sparkle! Suddenly, a friendly voice from inside the wardrobe said, "Hello, Timmy! I am the Magic Wardrobe. I can help you choose the perfect clothes for any occasion!"

Timmy was amazed. "Wow! Can you really help me?" he asked. The Magic Wardrobe nodded and said, "Of course! Let's start with casual clothes. These are comfortable and perfect for playing outside. Look, here’s a pair of cotton shorts and a tshirt. Cotton is light and keeps you cool."

Timmy quickly put on the cotton clothes and went out to play. He felt so comfortable that he ran faster and jumped higher than ever before! When he came back inside, the wardrobe spoke again, "Now, let's try something formal. These are clothes you wear for special occasions, like this neat shirt and trousers. They’re a bit more fancy."

Timmy tried on the formal clothes and looked at himself in the mirror. He felt like a grownup! "I look so smart!" he said with a smile.

Then, the Magic Wardrobe showed Timmy some clothes made for different seasons. "When it's cold, you need warm clothes like this woolen sweater and scarf. Wool keeps you warm and cozy," the wardrobe explained. Timmy put on the sweater and felt warm even though it was chilly outside.

"And when it's hot," the wardrobe continued, "light clothes like these cotton shorts and hats help you stay cool. For rainy days, you'll need a raincoat and boots to keep you dry."

Timmy was amazed at how each type of clothing had a special purpose. "But why do we wear clothes?" Timmy asked.

The Magic Wardrobe smiled, "Clothes protect us from the weather, keep us comfortable, and help us look nice. It's also important to choose the right clothes for the right weather so we can feel our best."

The wardrobe then taught Timmy another lesson. "Timmy, it's also very important to wear clean clothes. Clean clothes keep you healthy and make you feel fresh!"

Timmy nodded, understanding how important clothes were. He promised the Magic Wardrobe that he would always choose the right clothes for the weather and keep them clean.

From that day on, Timmy never had trouble deciding what to wear. He knew exactly what to pick, thanks to his new friend, the Magic Wardrobe.

## Activities

1. **DressUp Game:** Have the children bring in different types of clothes (casual, formal, and seasonal). Let them take turns dressing up in the different clothes and explaining why they chose that outfit.

2. **Clothes Sorting:** Provide pictures of different clothes and ask the children to sort them into categories: casual, formal, and seasonal. Discuss why each item fits into its category.

3. **Weather and Clothes Matching:** Show the children pictures of different weather (sunny, rainy, cold) and ask them to match the appropriate clothes to the weather shown.

4. **Materials Hunt:** Bring samples of cotton, wool, and silk for the children to touch and feel. Discuss the differences in texture and when each material is used.

5. **Clean Clothes Relay:** Set up a relay race where children need to "wash" (pretend with dolls or small clothes) and "dry" clothes, emphasizing the importance of clean clothes.

## Questions

1. What was the name of the boy who found the Magic Wardrobe?

a) Tommy

b) Timmy

c) Tony

d) Toby

2. What did the Magic Wardrobe help Timmy choose?

a) Toys

b) Clothes

c) Books

d) Games

3. What kind of clothes did Timmy wear to feel comfortable while playing?

a) Formal clothes

b) Casual clothes

c) Party clothes

d) School uniform

4. What are formal clothes used for?

a) Sleeping

b) Special occasions

c) Playing outside

d) Going to the beach

5. What material keeps you warm in the winter?

a) Cotton

b) Wool

c) Silk

d) Denim

6. What should you wear when it’s hot outside?

a) Woolen sweater

b) Raincoat

c) Cotton shorts and a hat

d) Winter coat

7. What kind of clothes do you wear when it's raining?

a) A swimsuit

b) A raincoat and boots

c) Shorts and sandals

d) A woolen scarf

8. Why do we wear clothes?

a) To look funny

b) To stay warm and protected

c) To play games

d) To hide from others

9. Why is it important to wear clean clothes?

a) To feel healthy and fresh

b) To play in the mud

c) To make them dirty again

d) To sleep better

10. What did Timmy learn from the Magic Wardrobe?

a) How to play games

b) How to choose the right clothes for the weather and keep them clean

c) How to read books

d) How to build a house

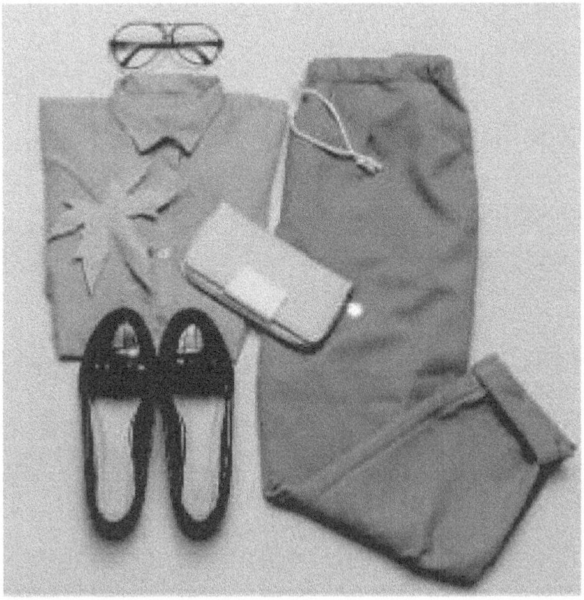

## Chapter 7 - My Family and School

Timmy was a happy little boy who loved spending time with his family. At home, he lived with his parents, his little sister, and his grandparents. Every morning, Timmy's mother would help him get ready for school, and his father would drive him there. His sister loved to play with him, and his grandparents always had fun stories to tell.

One day, Timmy asked his grandfather, "What is a family?" His grandfather smiled and said, "A family is a group of people who love and care for each other. Your parents, sister, and grandparents are all part of your family. We each have different roles that make our family special. Your mom and dad take care of you and help you learn new things, your sister is your playmate, and your grandparents share their wisdom and stories."

Timmy nodded, understanding how important each family member was.

After breakfast, Timmy went to school. His teacher, Mrs. Brown, welcomed him with a big smile. "Good morning, Timmy!" she said. "School is like a second family. Here, you learn with your friends and listen to your teachers."

Timmy loved school because he got to see his friends, learn new things, and play in the classroom. He knew that just like at home, there were rules to follow at school. Mrs. Brown taught them to listen to the teacher, be kind to their friends, and keep the classroom neat and tidy.

One day, Mrs. Brown asked the class, "Why do you think family and school are important?" Timmy raised his hand and said, "Family makes us feel loved and cared for, and school helps us learn and grow!"

Mrs. Brown smiled, "That's right, Timmy. Both family and school are important because they help us become the best we can be."

Timmy felt proud to be part of his family and school. He knew that both were special places where he was loved, cared for, and taught important lessons.

## Activities

1. **Family Role Play:** Have children act out different roles in a family, such as parents, siblings, and grandparents, to understand what each family member does.

2. **School Rules Chart:** Create a chart of school rules with the children, such as listening to the teacher, sharing with friends, and keeping the classroom clean.

3. **Draw Your Family:** Ask the children to draw a picture of their family, including all the members they live with. Let them share their drawings with the class.

4. **Friendship Circle:** In class, form a circle where children can share why their friends are important to them and how they help each other at school.

5. **School Scavenger Hunt:** Create a scavenger hunt in the classroom where children look for items that represent school life, such as books, pencils, or chalk.

## Questions

1. Who are the members of Timmy's family?

a) Parents, Sister, Grandparents

b) Friends and Neighbors

c) Teachers and Classmates

d) Pets and Toys

2. What does Timmy's father do every morning?

a) Plays with Timmy

b) Drives Timmy to school

c) Reads a book

d) Goes to work

3. Why is Timmy's family important to him?

a) They take care of him and help him learn

b) They buy him toys

c) They give him candy

d) They go on trips

4. What does Timmy love about school?

a) Seeing his friends and learning new things

b) Watching TV

c) Playing video games

d) Eating candy

5. Who teaches Timmy at school?

a) His grandparents

b) Mrs. Brown, his teacher

c) His sister

d) The family pet

6. What are some rules Timmy follows at school?

a) Listen to the teacher, be kind to friends, keep the classroom clean

b) Run in the halls, shout in class, play rough

c) Sleep during lessons, ignore others, leave toys everywhere

d) Bring candy to school, skip homework, play outside during class

7. Why is school important for Timmy?

a) It helps him learn and grow

b) It lets him sleep more

c) It gives him toys

d) It makes him stay home

8. What does Timmy's family do for him?

a) Make him feel loved and cared for

b) Ignore him

c) Only give him candy

d) Let him do whatever he wants

9. How does Timmy feel about being part of his family and school?

a) Proud and happy

b) Sad and lonely

c) Angry and upset

d) Bored and tired

10. What do family and school help Timmy become?

a) The best he can be

b) A superhero

c) A singer

d) A magician

# Lucy's Family Tree Adventure

Lucy was a curious little girl who loved asking questions. One day, she asked her mother, "Mom, what is a family tree?" Her mother smiled and said, "A family tree is like a picture that shows all the people in your family. It helps us see how everyone is connected."

Lucy was intrigued. "Can we make a family tree for our family?" she asked. Her mother nodded and got a big sheet of paper. "Let's start with you," her mother said. "You are at the bottom because you are the youngest."

Lucy drew a picture of herself at the bottom of the paper. "Now, draw your mom and dad above you," her mother instructed. Lucy carefully drew her parents and connected them to herself with lines. "This shows that we are your parents," her mother explained.

Next, they added Lucy's grandparents above her parents. "These are your grandparents they are the parents of your mom and dad," her mother said. Lucy added her grandparents to the family tree, connecting them with lines to her parents.

As they continued adding family members, Lucy began to understand how everyone was related. "So, Grandpa and Grandma are Mom's parents, and I am their granddaughter!" Lucy said with excitement.

"That's right, Lucy," her mother said. "A family tree helps us understand our family relationships and shows how we are all connected."

Lucy loved looking at the family tree they had created. "Can we add more people?" she asked. Her mother smiled and said, "Of course! You can add your cousins, aunts, uncles, and anyone else in our family."

Lucy spent the rest of the afternoon adding more family members to her tree. When she was finished, she proudly hung it on the wall. "Now I can see everyone in our family and how we are all connected!" she said with a big smile.

Her mother nodded and said, "Knowing your family is important, Lucy. It helps you understand where you come from and the special people who love you."

Lucy felt happy knowing she had such a big, loving family. She promised to always remember her family tree and the important connections it showed.

## Activities

1. **Create Your Family Tree:** Provide each child with paper and markers. Help them draw their own family tree, starting with themselves and adding parents, siblings, and grandparents.

2. **Family Tree Puzzle:** Create a puzzle of a family tree where children can piece together the different family members and connect them with lines.

3. **Family Role Play:** Have children act out different family members, like parents and grandparents, to understand their roles in the family.

4. **Family Picture Sharing:** Ask children to bring pictures of their family members to class and share them with their friends. They can talk about who each person is and their relationship to them.

5. **Family Tree Coloring Page:** Provide a coloring page with a basic family tree outline where children can color in and add names to the different branches.

## Questions

1. What is a family tree?

   a) A picture that shows all the people in your family

   b) A tree with leaves and branches

   c) A place to play games

   d) A storybook

2. Who was at the bottom of Lucy's family tree?

   a) Her grandparents

   b) Lucy, because she is the youngest

   c) Her parents

   d) Her dog

3. Who are the parents of Lucy's mom and dad?

   a) Lucy's grandparents

   b) Lucy's friends

   c) Lucy's cousins

   d) Lucy's toys

4. What did Lucy learn from creating her family tree?

   a) How everyone in her family is connected

   b) How to bake a cake

c) How to paint a picture

d) How to build a house

5. Where did Lucy hang her family tree?

a) On the wall

b) In the garden

c) In the kitchen

d) Under her bed

6. Why is knowing your family important?

a) It helps you understand where you come from and who loves you

b) It lets you eat more candy

c) It helps you play more games

d) It makes you taller

7. Who can you add to your family tree besides your parents and grandparents?

a) Cousins, aunts, and uncles

b) Toys and pets

c) Trees and flowers

d) Cars and bikes

8. What did Lucy's mother say about the family tree?

a) It helps us understand our family relationships

b) It is a place to hang pictures

c) It is a place to plant flowers

d) It is a game to play

9. How did Lucy feel after creating her family tree?

a) Happy and proud

b) Sad and lonely

c) Angry and upset

d) Scared and worried

10. What does a family tree help you remember?

a) The special people in your family and how they are connected

b) The names of all your toys

c) The places you have visited

d) The books you have read

Printed at Repro India Ltd.